The Fundamental Duality
of Nature

The Fundamental Duality of Nature

Richard Zindell

authorHOUSE®

AuthorHouse™
1663 Liberty Drive
Bloomington, IN 47403
www.authorhouse.com
Phone: 1-800-839-8640

© 2011 by Richard Zindell. All rights reserved.

No part of this book may be reproduced, stored in a retrieval system, or transmitted by any means without the written permission of the author.

First published by AuthorHouse   10/03/2011

ISBN: 978-1-4634-2072-7 (sc)
ISBN: 978-1-4634-2073-4 (ebk)

Library of Congress Control Number: 2011911130

Printed in the United States of America

Any people depicted in stock imagery provided by Thinkstock are models, and such images are being used for illustrative purposes only.
Certain stock imagery © Thinkstock.

This book is printed on acid-free paper.

Because of the dynamic nature of the Internet, any web addresses or links contained in this book may have changed since publication and may no longer be valid. The views expressed in this work are solely those of the author and do not necessarily reflect the views of the publisher, and the publisher hereby disclaims any responsibility for them.

Contents

Part 1 Something to Believe In 1

Part 2 What Nature Isn't 9

Part 3 Fundamental Duality 31

Part 4 Milky Way Lab 49

Dedication

I dedicate this paper to the people I care most about in this world. To my wife Barbie, my daughter Stefi, my daughter Tracy, and my son Scott.

Acknowledgement

A big thanks goes out to my daughter, Tracy, whose help was invaluable in the organization and preparation of this paper.

Part 1

Something to Believe In

It is the goal of many scientists to simplify the fundamental set of rules that nature uses to produce the full workings of our universe. They have come to be known as grand unified theories. Most end up attempting to join the four accepted forces of gravity, strong, weak, and electromagnetism. The success of merging three of the four forces at some point of temperature or time more than hints at their commonalities.

When Einstein was working on his initial theory of relativity, he came upon only two points of certainty that would end up shaping his views. The laws of physics are the same everywhere and the speed of light is constant. If these two simple ideas were true, then a great deal of the beliefs at that time needed to change.

If we were to craft a grand unified theory from scratch we would likely begin with any points of certainty and analyze the validity of any remaining beliefs. Those that are at odds with these basic certainties may need to be modified.

So are there concepts that have yet to be budged by the scientific method? The following four statements may be the right place to start.

1) **The conservation of mass/energy**—When Newton produced his second law of motion the calculations of all forces producing an acceleration on any and all mass was theoretically possible if one were equipped with enough information in two of the three areas. It was beginning to look like we lived in a deterministic world if we were supplied with enough figures and a good enough calculator. As we crossed into the twentieth century it became clear that Newton had pictured an errant, unchanging mass that led to larger and larger inaccuracies in an accelerating frame of reference. We came to understand that mass and energy were constantly exchanging identities and the combination of both were conserved. The total amount has never changed.

2) **Action/Reaction**—Newton's third law of motion stated that every action produced an equal and opposite reaction. The very discussion of grand unified theories is an attempt at reconciling the fundamental truths involved in forces always responsible for the production of actions. The validity of this law is currently in question with respect to an interpretive view involving a few specific areas within quantum

mechanics. Reasons for its listing as a concept to build upon will be discussed in Part 3.

3) **All quantitative language of the physical world is indeterminate at some magnification**—We've all used it at some point in our lives. We communicate our point of view and eventually a quantitative brushstroke creeps into our explanation. Just the most basic arithmetic as a way to make our issue concise. Addition, subtraction, multiplication, and division. These are the founding principles of applied mathematics.

Man had found a way to connect and equate seemingly disparate natural events and things with the juggling of these basic principles through the creation of equations. It all works nicely with just a minor exception. We can assign quantitative identities to just about anything. We just can't separate them into zero groups. This concept, a dividing by zero, is labeled undefined. You just don't do it. Newton and Leibniz found a brilliant way around this exception through the use of limits. We can approach any number, always getting closer but never reaching its undefined point.

Its use is not limited to a denominator of zero, but extends to our dealing with a singularity and the infinite as well. This is a coincidence that should not go unnoticed, as these are the identical triggers that prevent a successful merging of quantum physics and relativity.

The creation of a quantitative language has the same initial flaws today as when it was conceived so many years ago. It presumes that nature is in sync with an accurate assignment of numerical quantities to some natural qualitative equivalent. There is actually no such exact connection.

In manmade events, our system works well. We make the rules in determining a numerical equivalent. A touch of the plate by an offensive runner constitutes exactly one run. Nothing is lost in the translation from event to number. This is not the case outside man's rulebook of life. Any event, particle, or velocity assigned with a quantitative equivalent has an inaccuracy built in at its start. As we try to assign its initial number, we cannot pretend that a zero distance border exists. In other words, we are unable to know exactly where one measurement ends and another begins.

4) Reduction of all concepts must eventually be defined by space, matter, and their interactions—Parents have an ongoing dialogue with their children. A request is made to the child with a reply of why. Rephrasing the request in an attempt to make it more understandable just elicits the same response. The question eventually gets reworded to its most basic level.

Words as well as all forms of communication are an agreed upon assignment of meaning. If someone asks what you mean, you can substitute a different set of words for clarity. This can almost always occur because our language contains words that are not fundamental. That is, they are not the words that all others need be defined by. Two such words do exist; space and matter. They remain as the only fundamental concepts in our crossover from idea to the symbol of that idea.

This should be taken in its most literal form. In reality, these two words define our physical world at its most basic level. That would include the very forces that are responsible for the shaping of our universe.

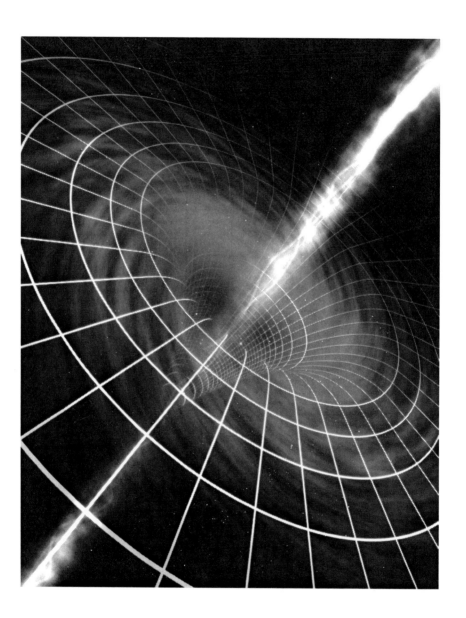

PART 2

What Nature Isn't

One of the best ways to look for inaccuracies in our present set of beliefs is a review of man's history in similar areas of thought. For our purposes scientific beliefs in varying periods works nicely. It isn't sufficient to just highlight what was once believed to be true. We need to look at why as well. Why would the intellectual societies of a particular period have such consensus and conviction on ideas that would so clearly be proven wrong? There should be good reasons for this in most, if not all instances.

We can find a treasure trove of discoveries throughout history that were once thought to be at the heart of the basic rules that are involved with or even governed over our universe. We just need to compare objective principles of today with the way our physiological senses had perceived these ideas when those were the best and sometimes only tools we had for scientific analysis. The contrast of these two perspectives tells a well known but compelling story.

Once upon a time, the experience of hot and cold would appear as fundamental aspects of nature. There is of course no such thing. Heat, or lack of it, is our body's sensation to the kinetic energy of molecules. It is movement of matter through space. Colors are difficult to describe, but certainly appeared to be intrinsic properties of matter. There is of course no such thing. All color is simply particular wavelengths in the electromagnetic spectrum. It is movement of matter's radiation through space. A body's apparent natural state of rest instead of the laws of inertia, the earth at the center of the universe with a boundaried flat surface, and the basic elements of earth, fire, air, and water. It only takes a few hours with any good book documenting the history of science to come up with hundreds of additional examples.

It may not seem all that important in light of the incredible technological advances we've made in the last hundred years. It is pretty clear that we have long since distanced ourselves from a reliance on our senses for an explanation of nature's inner secrets. It turns out that we remain dependent on something else far more unreliable than our own physiology for an accurate view on how our universe works. We are joined at the hip with the indeterminate aspects of our quantitative language.

One of the more interesting and ever-changing stories in scientific history is the classification of its players and their properties. Some were labeled particles and some were considered waves. The Standard Model holds the position that the zoo of particle structures in existence all exhibit both wave and particle tendencies that depend on the means of its measurements.

When Heisenberg formulated his "Uncertainty Principle", it became evident that man was unable to act as an omniscient character in the narration of nature's workings. The very act of measuring affected the outcome of its subject matter. What is not so well known was that Heisenberg initially named his results "Indeterminacy Principle". It may indeed prove to be a more compelling title.

Two of the major ideas in the Uncertainty Principle are:

$\Delta x \, \Delta p \approx h$

$\Delta E \, \Delta t \approx h$

The initial formula involves position (x) and momentum (MV $\rightarrow$ P). The minimum amount of uncertainty with respect to the product of both variables is never less than Plank's Constant (h). That minimum size becomes the basic multiple used in the quantizing of energy bundles. Those bundles are the basis of quantum mechanical description; the quanta exchange between matter and energy.

The second one listed involves energy (E) and time (t) and was a late addition to the uncertainty principle. After the introduction of special relativity, it was generally agreed upon that energy is to time as position is to momentum.

The increasing precise measure of one property comes with the cost of an increasing inaccuracy of the other. That's the uncertainty part. The indeterminate part is that none of the variables can be realized absolutely. This would leave the remaining variable as infinitely indeterminable. This concept filters into every attempt we make to classify and quantify nature.

The implications of both equations point toward quantitative limitations on particle confinement and borders. The upper limit of micro measure is also due to the undefined aspect of the instantaneous. Everything we've determined as particulate is in continuous interaction and exchange with its surroundings. Nature is a movie with no pause or stop button.

The possible states of any given system or area of measure require wave equations for accurate description. It serves as a probability mapping of position and momentum with every test particle's measure acting as an overlay on previous results. Only in bypassing and stopping these exchanges as a mental notation through measurement can we collapse the wave function and produce the nomenclature of particle en masse.

It remains likely that the first few decades of the 21st century will be home to a number of fundamental beliefs ripe for dismissal by coming generations. One likely candidate is time.

Time is more often viewed today by its apparent flow than by endowment of a set of intrinsic properties. It is usually described as flowing in one direction. Before dealing with this view head on, we need to review another likely candidate ready for discarding as a fundamental property of nature. I am referring to entropy. A quantum review of entropy will provide a much clearer basis for the rationale of finding the idea of the unidirectional flow of time unfounded.

Entropy is the second law of thermodynamics. It is often viewed synonymously with time's "one directional flow". It can be described in two basic ways:

1) It is the constraint on the direction of heat transfer. It limits the amount of work generated from X amount of heat to be less than 100%.

2) The total disorder in a system is always increasing. A viewing of two scenarios that occurred separately would allow a viewer to construct a timeline. Unlike the world of classical physics, entropy allows a determination of the more disorderly scenario as the more recent of the two. It should be noted that the law allows for external contribution of order to a system providing that it is more than offset by the increasing entropy of the whole.

Both descriptions run along parallel lines of logic and intersect within a broad enough interpretive view. If the universe has always been preceded from an ever greater order, then the flow of time appears to take on meaning. The premise this conclusion is based on though is completely false. A closer view at the micro level tells a far different story. It turns out that while it is clear entropy is the favored path of nature, it is by no means an exclusive one.

Subsequent to the discovery of Brownian motion and quantum mechanics, the random irregular motion of atomic and subatomic particles was as real as any macro movements we bare witness to each day. A particle's velocity and direction of motion is not solely determined by energy sources of known origin. The overwhelming majority of gas particles in passive transport end up diffusing through any open space. Eventually, all moving particles display a tendency to spread from heavier concentration to an area less so. From a quantum perspective, each individual particle's area of distribution is defined as the probability or likelihood of being found at a particular point. An absolute determination is made only when measured. Prior to that nothing is actually defined at zero or 100%.

In our example, the probability curve is partly based on the amount of space available. Simply put, crowded areas have less open spaces for travel and position. This is exactly what occurs in the human world. Every concept man has gone through the trouble of naming comes with an almost infinite amount of neighbors. The neighbors remain unnamed, but exist as a sea of alternative outcomes competing with the one and only named state in our scenario.

Some of the most basic concepts of probability and chance are taught to grade school children by way of coin tosses and playing cards. For those adults who missed that lesson, they need only find an open seat at their favorite poker table.

In the early days of poker, each player would receive five cards as their full playing hand. If your probability skills were up to par, you could calculate the odds of receiving any combination of those five cards. For those whose interests lied elsewhere, cards listing the odds of receiving each poker hand were available for purchase.

There are a little over two and a half million possible five card combinations arrived at by 52 factorial. A royal flush (poker's best hand) can be expected four times per two and a half million hands since there are four of them. While sitting at a game with a friend, he draws a royal flush and you draw a two, five, and nine of clubs and a jack and queen of hearts. You discuss the odds of getting a royal flush versus the identical hand you just received. Your basic training in probability proves that it is four times more likely to draw a heralded royal flush than duplicating your particular hand.

The point, of course is that every individual combination of cards carries the same two and a half million to one odds against drawing it. The hand is dismissed as common because the game's rules bundles more than 90 percent of the combinations as likely losing hands if played to the end. Every trash hand becomes synonymous with all others. They are the sea of alternative outcomes competing with man's preferred state. Every low probability of outcome aside from the very formation of life is created out of man's use of naming and classification. They are not naturally derived.

Nothing is actually preventing an orderly state outcome. Nothing is actually preventing the random unidirectional movement of all subatomic particles in a person's body to produce flight and a seeming defiance of gravity. Quantum mechanics allows for a net decrease of total entropy while probability inhibits the likelihood of its occurrence.

When Einstein paired time as a kind of fourth dimension along with the positional notation of matter in dimensions x, y, and z, he demolished the static and absolute pedestal time had always been placed upon.

Time was now bundled with space, preventing its full understanding and significance as a standalone concept. But what really is it within nature's laboratory? Newton's second law of motion, $F = ma$, brought us one view. Acceleration is the derivative of velocity. Getting time by itself gives us $t = mv/F$.

Time then can be viewed by integrating each small measure of momentum per unit of force. It should be clear that defining a fundamental concept by way of a collection of uncertain magnitudes of motion may not be a wise path to take. It also serves as a rate of change in periods of seconds that remain as a relative comparison to other physical events such as the rotation of earth around our sun.

In Einstein's view of x, y, z, t it appears as the bookmarking of location or event of a point particle at some distance from the origin of measure in width (x), height (y), and depth (z) at some relative notion of arrangement of a given area's other molecular structures. In fact, the t in any and all instances requires a complete cessation of movement by all universal matter, even at the quantum level. Time turns out to be a convenience for most biological entities. It remains an indispensable measure to man but totally constructed by piggybacking on other non-fundamental interactions of nature.

Its tethered link with space can be seen in Einstein Minkowski Space-Time diagrams. A simplified version follows with full attention given to any fundamental role time takes on in the interpretation of travel through spacetime.

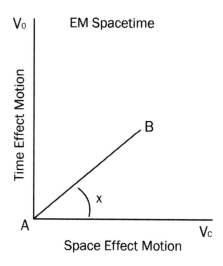

A particle's motion along line AB is plotted somewhere between line of time at velocity Vo (resting) with respect to A and line of space traveling at Vc (lightspeed) with respect to A. One of the more prolific uses of this spacetime diagram compares the aging process of particulate matter along the line of AB. Matter in motion approaching B ages more slowly than matter at rest in the vicinity of A. This is a graphic illustration of approaching the complete dilation of time as the cosine of x approaches one.

Another simple drawing should shed light on the actual variables involved.

The Fundamental Duality of Nature

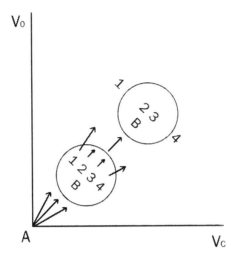

Particulate A is at rest and particulate B is in motion through spacetime with respect to A. Particulate B is comprised of subunits 1, 2, 3, and 4. As B approaches Vc, 1 and 4 separate from B as their line of motion is not consistent with its remaining subunits 2 and 3. As B reaches Vc, subunits 2 and 3 do as well. B's retention of Vc, with respect to A will ensure that only subunits still identified within B cannot be undergoing any change of arrangement between themselves. Subunits 2 and 3 remain in a state of identical positioning with respect to each other. There is no change of motion hence no aging. Here again, the notion of time and any interpretation of its dilation is nothing more than an illusory principle. Time remains a fabricated concept that can only be created from or patterned after a targeted, natural event.

If time is removed as a tool of measurement, certain quantities calibrated to its change needs to be addressed. The speed of light is quantified in miles per second. Is the utilization of velocity in the comparison of two or more motions accurate at all, since it relies on non-fundamental data? Is there any meaning to the upper limits tied to the electromagnetic spectrum?

Our quantitative language creates difficulties in the resolution of these issues. There are ways to visualize an upper limit of speed without an exact measure. Our graph above shows particle B and its subunits 2 and 3 approaching Vc. In an effort to increase its speed, the net increase of applied force would have to equal its change in momentum (MV). The work added would be identical to its change in energy. That required change would be infinite. In other words as $V \approx c$ then $E \approx \infty$. The upper limit of velocity is realized through meaning rather than absolute measure.

Upper limit and absolute measures involving mass and length take on meaning when viewed from an inability to change any further. This is clearly seen in particle B's approach to Vc. The increasing applied force on units 2 and 3 collapses any spatial barrier between all particulate matter. Its mass per unit of volume can go no higher and it length can contract no further.

Time is also a reference to nature's tense of past, present, and future. Matter is in continuous rearrangement. The future is nothing more than an expectation of matter arrangement in some patch of locality. The present is the combination of our movement through a local part of space and matter and some part of space and matter's movement past us.

The ongoing exchange of space and matter presents a number of alternative ways to bundle up a particular group of these interactions for naming and classification. Living entities can utilize any of their senses to border various sets of interactions to become virtually any verb or noun of description. Once done, nature's exchange is halted and a newly created whole has been conceived. History and all past tense is the accumulation of these created states.

Quantum mechanics defines a process known as quantum decoherence that connects the measure of nature's activity through a series of wave equations. The equations produce only the probabilities of a physical occurrence. When a measurement is made, the wave equations collapse creating a single recognized state. One interpretation of its significance is a past that is created only through measurement.

A number of theories have come to the forefront of science quite recently. Many are conceived to address numerical imbalances within the four recognized forces of nature. Dimensional and multiverse analysis are two such attempts at reconciling microphysical experimentation and theory.

Why would such an unusual set of constants be necessary for the universe to exist in the first place? How is nature able to adopt an almost exact combination of parameters to allow life to initiate and flourish? An additional group of hidden dimensions and universes would allow all possible arrangements and outcomes to exist. That which we are experiencing is just one of many.

The manipulation of the universal accounting ledger by way of the creation of the multiverse is philosophical semantics. It can neither be directly proven or measured. The number of dimensions contained within the universe is a different matter.

String Theory is a mathematical gem that equates particulate matter with the geometry of vibrating strings. The added dimensions allow all the numbers to balance. Much of string theory's present format necessitates the use of 10 or 11 dimensions to account for the full array of particles now thought to exist within the standard model. Most audiences experience some level of difficulty in visualizing these extra dimensions. They are usually given an example of compactification by way of the long distance viewing of a garden hose. The length is clearly seen while its width and depth may escape notice. Only when viewed up close or by higher magnification will all three dimensions be apparent.

Throughout recent years, the accuracy of measurement has grown substantially in both the micro and macro arena. There is absolutely no evidence that nature hides its dimensions. It does quite the contrary, as each spatial dimension creates an outward direction in all magnitudes of measure. The fundamental building blocks of nature are within our ability to analyze. The language we end up using to describe it may not be adequate.

Our quantitative expressions must allow an accurate entry of all variables and constants into every equation describing nature. At present, a singularity and the infinite cannot be adequately represented as their quantitative existence remains undefined. String Theory succeeds in avoiding this trap by defining all existence in multiples of strings never smaller than Plank length.

String theory can be viewed as a brilliant line of mathematical reasoning created to fit the evidence as it unfolds. It can't actually be wrong since it changes to ensure that its latest version remains quantitatively in sync with nature. Its flaws will continue to be two fold. It has not been, nor is ever likely to be, verified. More importantly, it is likely to be one of a number of quantitative ways to explain scientific phenomenon. It is as if a teacher gives her pupils a description of a situation and the task is to find the scenario that would explain every aspect of that situation. The right group of students will end up with a number of equitable solutions.

General relativity takes a conceptual approach by endowing the universe with positive spatial curvature. This allows the inference of finite volume without edge. Geometric examples are ubiquitous. They all of course reside within the existing whole. A crossover to represent that whole by a boundaried curvature may be described in words but it fails in any numerical equivalent.

Quantum mechanics endures as the most successful approach in quantifying the particle/wave interactions of matter and energy. Some of its support came about through a reactive attempt to solve the problem of blackbody radiation. The electromagnetic spectrum was viewed as a continuum of escalating frequencies from long to short wavelengths. A perfect black body was believed to radiate the energy it could no longer hold throughout the frequency spectrum. This would lead to an infinite release of energy in the smaller wavelengths part of the curve where the highest number of frequency modes exist. That, of course, was an absurd answer to be left with. It was proposed that energy states of a physical system must be discreet. All radiated and absorbed energy must come in bundles or quanta based on the product of its frequency and Plank's constant.

In effect, the constant serves as the smallest and most fundamental connection between quantum wavelength and every particulate of matter known or suspected to exist. Through the combination of different variable equations, Plank's constant can be used to derive the base unit of almost any physical quantity including length, volume, momentum, and force. Through the use of wave equations, uncertainty, and Plank's constant, force and matter calculations are maintained well within the lines between the singularity and the infinite.

I am certain that most of the great scientific minds throughout history would be astonished at the body of knowledge we have at our disposal. Our existence in a slow moving macro world is clearly not representative of how the universe transacts its affairs.

Much of what is taught today is both brilliant and practical, as evidenced in many ways. For the most part, it is also invented. Our entire mathematical system along with our generally accepted notion of time are constructs of man. This does not make them any less of an achievement. On the contrary, they may in fact be the most cherished and utilized creations of mankind.

Space and matter were evident to hominids of very early decent. They also remain as the most fundamental discoveries that exist whether or not any mind is around to experience them.

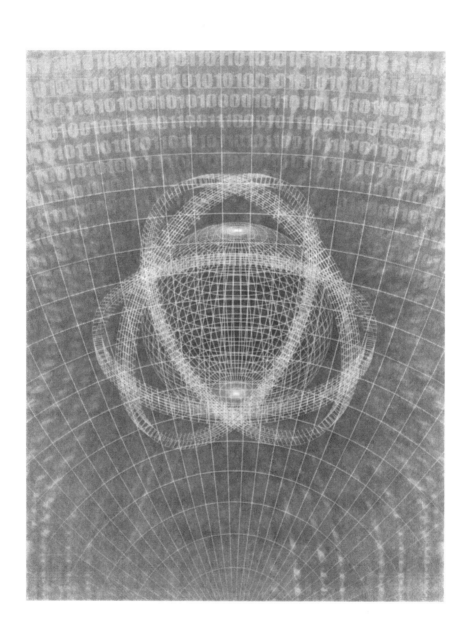

Part 3

Fundamental Duality

Richard Zindell

Is everything we've come to know just an indexing and calibration to other words that never fundamentally address nature? Is there a difference between a description of nature that caters to our physiological makeup and one that objectively treats the universe only on its terms? What does nature actually engage in regardless of whether or not there is life to interpret such actions? These questions are not answered by running man's time clock in reverse to arrive at a consolidation of mass/energy prior to its big bang explosion. The answer is movement and interaction with the only two things nature has at its disposal. Its matter and its space. All of our concepts, interpretations, and descriptions start form here. Everything we know is eventually defined by this.

Any search for the creation or appearance of nature's fundamental building blocks straddles the gateway between scientific method and spiritual leanings. We will not be crossing that path in this paper. It would follow then that all conceptual ratios and equations can be simplified or reduced to its fundamental alignment of Space/Matter (SM) interactions: work and power, velocity and acceleration, energy and even force. The four fundamental forces should just turn out to be subsets or offspring of its SM origins. This is the proposal being set forth in these writings.

It has been evident for some time that Newton's first law of motion F = ma is valid in an inertial frame of reference only. His third law of motion also came into question as it clashed with quantum mechanics in a few specific areas. The law states that the force between two bodies is always equal and opposite. The conservation of momentum can be derived from this since the net exchange of momentum is zero. This specific use of the law has always been upheld.

The general interpretation of the law appears to take exception with two areas of quantum physics:

1) Momentum of force fields

2) Quantum entanglement

Newton's third law, which can be referred to as action/reaction, goes to the heart of SM interaction. We will need to take a closer look at both areas listed above. The reason is not to confirm their existence, as has been done on many occasions, but to ensure its interpretation is consistent with nature's workings.

The first exception involves the momentum of field measure. Force fields represent the potential exchange between a point particle and its surroundings. A changing field is the geometric rendition of spatial change. It is a visualization of the shift in momentum particulate matter will undergo in every incremental change of its position.

Electromagnetism is actually a subset of SM interaction. The electromagnetic field has no boundaries. The current producing a magnetic field is simply the orderly flow of electrons. In a more general sense, it is the movement of matter through space. The changing magnetic field is an accurate portrayal of spatial's altered potential.

Fields propagate at lightspeed. Charged particles attract and repel its matter neighbors at a snail's pace. This large difference of magnitude stems from the idea that SM interactions are fundamental. Matter and matter reactions are not. The latter is governed by chemical processes and not by nature's basic exchange. This basic exchange was exactly what Maxwell discovered when he came upon the realization that the speed of light was identical to any wavelength in the electromagnetic spectrum. This is nature's tempo for all SM exchange.

Maxwell brought together four equations from Gauss, Faraday, and Ampere to describe and quantify the connection between electricity and magnetism. These equations, as well as all field measures, use line and surface integrals to sum up interaction strengths from each small area of measure. The use of integrals creates infinitesimal gaps of inaccurate measure that can add up to a skewed interpretation of results. It can also serve as a disconnect between reactions brought about by actions and those of random origin. This potentially errant process can be seen quite clearly by a review of integral calculus's most basic equation.

The definite integral of f along the closed interval of |a,b| is given by:

$$\int_a^b f(x)\, d(x) = F(b) - F(a)$$

We will turn again to the inaccurate use of boundaried entities and positional points to see where potential problems may arise. The inference any proper integral makes is the exact measure of its integrands |a,b|. The results of interval measurement is errant not by calculation but by the conclusions drawn from it. Vector calculus and the use of a dimensional Laplacian operator runs into similar problems. All measurements utilizing dimensional positioning within spatial intervals produces misleading entity states.

Zero distance bordered intervals remain undefined. This does not create difficulties in relative measure. The exact value of each variable is less important than knowing how they relate to each other. A complete measure of SM border exchange is never just an exercise in relative valuation and change. Wave equation measurement within any SM border exchange is always impacted by an undefined quantization of the instantaneous and zero distance borders. The random nature of any entity's momentum arises from a flawed interpretation of quantum mechanical calculations.

If your interests tend toward the bizarre, you are likely to find the study of quantum entanglement rather appealing. When particles interact, they take on an extended set of shared identities in a quantum mechanical description of the pair. They can be viewed as a list of complementary measures in the areas of momentum, position, and spin. A counter clockwise measure of spin on one particle would render a clockwise spin on any future measure of its partner.

Any pairing of properties remains in an undefined state until a measure occurs. The interpretation of connection is unclear because the remaining entity always takes on the complimentary value regardless of the distance separating both particles.

The rationale for this "action at a distance" has been debated for many years. Our incomplete understanding of the process fails to lend any substantive support for any lack of action/reaction involvement. The initial act of measure is not likely the only variable involved, but it has proven to be a necessary action to produce its complimentary value in every case.

The universe covets its total supply of mass/energy and momentum/angular momentum by pairing all actions with an equal summation of reactions. This then begs analysis from the results of spatial reaction to matter's action and matter's reaction to spatial action. The integration of these actions should serve to quantize all subsequent events. In short, matter curves space and the action of changing acceleration of matter reacts as spatial energy storage. Matter's contact with a changed spatial state affects the motion of that matter. This is the ongoing dance of nature and serves as the backdrop for all that currently exist.

The ratio of surface area to volume plays a prominent role in the world of biology. The surface of entity states generally serves as a means of exposure and interaction with its surroundings. An increasing ratio provides for a greater exchange with its interior. This is essentially what occurs with spatial curvature and matter. Higher ratios give rise to a more concentrated SM exchange. These effects along with their directional netting will help to bring about common ground for the unification of the four forces of nature.

Beginning from the micro level, this set of interactions forms the basis of what we've come to know as the strong force in the nucleus of the atom, neutron stars, and black holes. In all three cases, matter is being contained in smaller and smaller boundaries by a previously changed spatial state. The difference in magnitudes arises from the varying ratio of curved space per unit of particulate matter.

The strong force is viewed as the securing strength necessary for the proton's constituent parts to remain together. This ability of quarks and gluons to both exchange and remain stable is represented by a force that does not diminish in strength with respect to increasing distance. It is believed that the residual effects of the strong force produces the nuclear force, which allows normally repelling protons to stay confined within the atomic nucleus along with any neutrons present. This force acting on nucleons, does diminish in strength as the distance between particles gets beyond the length of its own diameter. Within this distance, force particles known as mesons are exchanged to offset the repelling nature of like charged particles.

The strong force is measured to be 10^2 times that of electromagnetism, 10^{13} times the weak force, and 10^{39} times the force of gravity. At first glance, the difference in magnitudes of strength between any of these forces seems both colossal and arbitrary. Science has, of course, been able to unify three of these four forces under extreme experimental conditions. The lone holdout continues to be gravity. It turns out that all four forces are really the residual effects of SM interactions. In order to see why, it is best to take a close look at both the magnitude of the strong force and an explanation of the what and why of gravity. The combination of these two areas goes at the heart of all SM activity.

When Newton created his laws of motion he never attempted to produce a graphic image of gravity at work. Einstein's efforts at describing and actually visualizing the warping of spacetime gave many people that highly sought after "picture worth a thousand words". Highschool physics books portrayed a textured fabric stretched tight with a bowling ball's weight producing a curvature on the fabric previously not there. A lighter ball placed at the outside of this fabric rolled along a newly created path with the slope and velocity dependent on the mass of the bowling ball. This was the picture of mass induced curved space. The rotation of all heavenly objects was due to a larger body's warping of space somewhere close to the center of its plane of revolution.

The force is calculated by a relatively simple equation.

$$F = G\, m_1\, m_2\, /\, r_2$$

The gravitational attraction for any two particles is the product of their masses with a gravitational constant, varying inversely with the square of its distance. But there's a catch. For any two test particles it works well. But the entire universe is filled with ongoing matter/energy exchange. In any spherical, three dimensional test area there are an astronomical number of matter interactions. Some are constructive to a particular test measure, but the vast majority are destructive to any line integral connecting each test particle's center. The aggregate effect of all gravitational actions in any body of matter greatly diminishes any cohesive unidirectional pull on matter. The result is the appearance of a relatively weak force.

These results occur because of our titling of entities. The earth is one body while the moon and sun are two others. This is not to argue that the earth, sun, and moon are not real. It is just that nature is in the business of SM interactions.

Any collective entity takes on an average of the myriad of fundamental interactions that occur within its boundaries.

The origin of all magnitudes of motion and the dilution resulting from their combined measure goes beyond any quantitative inaccuracies. It creates the illusion of intrinsic properties of spin to virtually all particles and bodies of matter subjected to testing. This is not easy to describe in words, but the general idea should come through.

The shape and boundary given to any entity is usually derived from the reflected portion of some segment of the electromagnetic spectrum. In the macro world it is usually visible light. When we visualize a magnet, we sometimes add a mind's eye view to include field lines connecting the north and south poles. Most people would not find controversy with the use of either pictorial. The first picture is the physical magnet. The second also includes the direction and strength its field reaches outside the borders of reflected visible light.

What about a picture of our solar system? That's considered a whole entity. The outer planets are connected to the sun by a gravitationally induced rotation. But what about any varying thickness of this entity across the plane of Neptune's orbit around the sun? Field forces generated from the gravitational pull of other planets outside its plane of rotation will create a typography of force sure to be considered as part of the makeup of our solar system. We end up with a drawing not unlike a 1950's Hollywood version of a flying saucer. The point of course is the infinite amount of ways we can utilize subjectivity with some amount of rationale and conviction to produce a rendering of any physical object. The very act of utilizing anything but nature's fundamental units to determine entity states allows for symbolic meaning only.

When we endow intrinsic properties of spin to any titled entity, we have created an arbitrary whole with a center of gravity based on a distribution of the myriad of particulate matter within its boundaries. Rotation is integral to the motion of every particle of matter in the universe. Each particle is compromised by both its inertial line of motion and the tug of attraction to coiled space.

Our view of any collective of particles as one entity state gives the appearance of spin. Similar velocities in a common locality are generally the criteria that mold so well to our view of a cohesive unit. It is important to distinguish this description of motion from the covalent, ionic, and hydrogen bonding involved in the shaping of matter. Particles do combine to make larger entities and they do spin. It just doesn't occur within nature's fundamental exchange.

Our picture thus far envisions matter's effect on space. As every action produces an equal and opposite reaction, we must analyze spatial reaction which has clearly been acted upon by matter. Our mind's eye picture of gravity has our sun warping spacetime in a relatively close section of our solar system's center. We need to extend this pictured view to include a moving solar system within our galaxy and a moving galaxy within our universe.

Passing matter that affected space has moved on. The changing acceleration of rotational bodies has caused a shift in mass/energy. But a shift to where? The answer is in the changed state of space left by any passing matter. The best way to quantify this newly coiled space is to see its effect on any matter passing through it subsequent to its change. All this is happening at the macro level with stars and planets as well as at the micro level with protons and quarks. It is also happening everywhere in between including atoms, molecules, and all other particle entities.

A review of the magnitudes of the strong and nuclear force gets us much closer to discreet gravitational or SM interactions less compromised by larger bodies. We are still unable to arrive at exact strengths since any distinction of particle identification assumes zero distance borders. Our current mathematical language allows only wave equations and probability curves as our most accurate view of nature. Man's nomenclature of anything is synonymous with its measurement. It is only at this juncture where there is a collapse of the wave equation, that allows us to arrive at a particle-like solution. It is never answered in the present. Past tense is determined from its recognition and measurement.

Science has the current force count at four. Some would say two if you consider partial unification. For our purposes, no actual number is used. We'll just say that all universal activity stems from SM interactions. Any truth to this view necessitates or minimally hints at some intervening exchange. This is where we need to initiate a closer look at energy.

Science defines energy a number of different ways. The two most generally used are:

1) The ability to do work where work is a line integral of force along path c (as noted below).

$$W = \int_c F \cdot ds$$

This can be summarized as force through a distance.

2) $\frac{1}{2} mv^2 + mg \Delta H$

This can be broken down as $\frac{1}{2} mv^2$ representing the kinetic or moving portion while $mg \Delta H$ is its potential. It is evident that both equations allow more for measurement and magnitude than description and meaning.

Energy should be viewed as the basic exchange between space and matter. When science defines it as a combination of kinetic and potential, it is from the vantage point of matter. This is understandable since energy's measurement is a force placed on matter through a distance. Since nature actually works through SM exchange, it stands to reason that matter's kinetic energy is transferred as spatial potential. There is nothing inherently preferential in viewing energy transfer from the perspective of matter. Energy exchange has no preferred frame of reference and is equally valid from either perspective.

Richard Zindell

The quantitative universe is like a set of checking accounts in the name of space and matter. The kinetic action of one debits its balance as a check written. The receiving account deposits it as potential for future kinetic action. The quantities swapped are identical. The total balance of mass/energy is conserved.

PART 4

Milky Way Lab

Attempts to unify the fundamental forces of nature is a work in progress. Relativity creates an effective visualization of moving bodies that appeals to the rational mind. Quantum mechanics continues to provide predictive powers over natural events that remain less than fully understood. They may both claim to provide conceptual provisions for a singularity of origin or measure, but neither has the equations equipped to back it up.

Regardless of any measure of trust one has in the current set of theories governing the laws of nature, one thing remains certain. There are areas today that remain unsolved and many would agree that this will always be the case.

We've arrived at a crossroads in the pursuit of alternative solutions in lieu of unsatisfactory explanations. There are a number of prominent unknowns in the world of physics today. We need to see if nature's most fundamental interactions are capable of shedding more light on areas so in need of clarity. It would be far better if we could find a way to test what has been presented here and figure out what the results tell us. For this kind of task we'll be needing a big laboratory. Let's make that a really big laboratory.

The world of science is divided on a number of issues. Although subjectivity certainly plays a role in deciding which are worth looking into, the following four areas of study remain at the top of many physicists' worklists.

1) Increased rate of expansion by accelerating galactic motion.

2) Dark energy, the proposed solution for #1

3) Accelerated rate of rotation by star systems in the outer galactic region.

4) Dark matter, the proposed solution for #3

The discovery of Doppler redshift and its subsequent connection to distances of galaxies clearly evidenced an expanding universe. The idea of running the motion in reverse and arriving at the initial point of expansion was the path science would take to determine universal momentum. A recent problem was discovered. The universe is expanding at an accelerating rate. If all the mass/energy was confined within the big bang, where is the additional mass/energy necessary to fuel this accelerated rate? The proposed solution comes from dark energy, which is now believed to make up 73% of the total mass/energy of the universe. The number is arrived at by calculating how much we need to explain current measurements. It is homogenous and acts through gravity, but nothing else is known about it.

In the 1970's, a pair of astronomers were studying the structural differences of spiral galaxies. A part of the study warranted the comparison of movements of stars near the galactic center with those on the outer reaches. The results evidencing equal or greater velocity of stars farther away from a heavier center of mass was anything but expected. The proposed solution comes from dark matter, which is supposed to make up 23% of the total amount of mass/energy in the universe. It is currently undetectable.

Just like string theory and maybe even more so, dark energy and dark matter are forced solutions provided to fit newly found data. The percentages attributed to their existence are backed in numbers. They are another version of cosmological constants and not well disguised at that.

Before we jump to fancy and fabricated possibilities to fit the facts, it may be helpful to start at the origin of nature's real interactions and see if there are better hints there. At first glance, the meager descriptional differences of dark matter and dark energy seem to warrant the classification of inverse forces. After all, an accelerating universe appears to be a repulsive force acting antagonistic to gravity. Evidence of faster rotation from stars in the outer galaxy seems to suggest an attractive force consistent with gravity.

There is another answer that is completely consistent with everything nature is already doing. The action/reaction of space and matter leaves a potential energy that is shared with all subsequent passing matter. This changed pathway produces an attractive force within the affected spatial field, no different than the gravity we experience every day. This interactive exchange is likely the equivalent of the proposed Higgs field in that it is the exchange of mass intrinsic to the quantification of matter. Unlike some descriptions though, it is not a field endowed with an equivalence of measure in all directions. This explains why dark energy needs to be almost ¾ of the mass/energy of the universe. This is because the scientific community structured the dark energy density to be present in all volumes of space.

But what about proof? How could we design an experiment to verify the results of SM interactions? The obvious answer would be to send a substantive body of particulate matter beyond earth's orbital. We need to follow the path it generates with another substantive body and see if this second body is impacted by an accelerating force. It doesn't seem likely to happen, in that the largest object we send into orbit is a relatively small rocket ship with nothing in its wake. We may not have a manual lab at our disposal, but we certainly have one from nature that is far bigger and better. We have got our own Milky Way galaxy.

Planetary revolution was calculated by Kepler and explained first by Newton and then by Einstein. The combination of the sun's gravitational field and its inertial motion compromise as orbital rotation. The large proportion of mass in our solar system's center ends up warping the spatial path there more than at its edge. Mercury and Mars have higher rotational velocities than Neptune and Uranus.

As mentioned earlier, star rotation in the galaxy was now shown inconsistent with what we knew to be true in our solar system. But there is another big difference in the two. The planets in our solar system are not generally compromised on a regular basis by other large bodies in its rotational path. The Milky Way presents a far different story. There are a myriad of star systems in any given revolutionary path around the galactic center. Every star is essentially in tow with those preceding its journey. It is precisely the enactment of the experiment we needed to build.

A changed spatial state left in the wake of a moving star system alters the path of any passing matter. This is similar to the altered path starlight takes as it passes the changed spatial state near our sun. Solar system rotation is a special case. The accelerated change of matter following rotating star systems would not come in the form of directional shift since both lines of motion are in sync.

A reacting spatial field would produce an accelerating attractive force between both system's center of mass. Its inertial state would be shifted toward the tangential. The integration of spatial recoil from others in its path more than offsets the large distance from its crowded center. It is SM interaction and not a fabricated, mysterious, dark matter force that produces a higher than expected rotational acceleration of the outlying star systems in our galaxy.

The expansion rate of the universe has been subject to many tests over the past three decades and the evidence clearly shows it is expanding at an accelerating rate. Running this expansion in reverse brings us to an extremely compacted state of matter. The circumferential exchange of matter was at its highest density. The outward expansion created changed spatial states found in every path subsequent passing matter would encounter. Every celestial body engages in SM exchange, leaving an effect on its spatial surroundings. These are the pathways that shorten matter's journey.

Matter's accelerated motion would be driven by coiled spatial areas in its path of travel. The identical attractive force shortening pathways of rotating star systems is also at work pulling at entire galaxies and quickening their pace. SM interactions don't just explain this. It predicts its occurrence.

Extrapolations taken from the study of vacuum energy and quantum field theory point toward a substantial collective energy potential within space. These theories need to take the next logical step involving the recognition and description of the ongoing exchange of that potential with its matter neighbors. The constructive pursuit of an accurate accounting of spatial potential will deliver the additional matter/energy science has been searching for. Balance will be achieved through the full use of the real variables of nature.

So there it is. A series of SM interactions leading to a complicated universe. Nature's story is not unlike our own in the utilization of simple binary code to create the full array of computer applications we marvel at today. Nature uses the only two arrows in its quiver. Such splendor from humble origins.